Understan... Sound and Light

Printed in Mexico

ISBN-13: 978-0-15-362257-1
ISBN-10: 0-15-362257-1

2 3 4 5 6 7 8 9 10   050   16 15 14 13 12 11 10 09 08

Harcourt
SCHOOL PUBLISHERS

Visit *The Learning Site!*
www.harcourtschool.com

What Is Sound?

How do instruments in an orchestra produce such beautiful sounds? Each instrument—drum, guitar, flute—produces a different sound. Together, the sounds combine to make beautiful music.

Sound is a form of energy that travels through the air. You probably know that different musical instruments not only make different sounds, but also make sounds in different ways. A drum makes sound when it is hit. A guitar makes sound when it is strummed. A flute makes sound when air blown over the hole in the side of the flute causes the air inside the flute to vibrate. However they are played, all musical instruments make sound when something— the head of a drum, the strings of a guitar, or the air inside a flute—vibrates. A **vibration** is a back-and-forth movement of matter.

Most of the time, the sound you hear travels through the air. But vibrations can travel through any phase of matter—solid, liquid, or gas. Matter that carries sound waves is called a *medium.* The speed of sound depends on the medium through which it is moving. All sounds travel through a certain kind of medium at the same speed. If the medium changes, so does the speed of the sound. Sound moves faster in warm air than cold air. Sound also travels faster in solids and liquids than it does in gases.

The sounds of a flute are produced by a vibrating column of air.

 MAIN IDEA AND DETAILS
Describe three ways musical instruments produce sound.

Whales produce sounds that travel under water for hundreds of kilometers.

Fast Fact

Speed of Sound Through Different Mediums

Medium	Speed of Sound
Air (gas)	340 meters per second (about 1100 feet per second)
Water (liquid)	1500 meters per second (about 4900 feet per second)
Steel (solid)	5000 meters per second (about 16,400 feet per second)

How Sound Travels

Sound travels in waves. When the head of a drum is struck, it begins to vibrate. As it vibrates, it pushes on the air directly above it. As the sound travels through the air, molecules in the air vibrate in the direction the sound is moving. As the sound moves, the molecules of air are compressed, or squeezed, together. The compressed air pushes on the air next to it. The vibration continues to spread. The compression is passed along, producing a series of sound waves. Sound waves move out in all directions from the object that makes the sound.

Sound travels in compression waves. The waves move out in all directions.

When sound waves hit a surface, some or all of the sound energy is absorbed by the surface. Soft surfaces absorb more energy. Hard surfaces absorb less energy, so a sound that hits a hard surface bounces back. **Absorption** is the taking in of light or sound energy by an object.

HELLO))) (((hello

When you shout, you hear the sound as soon as you make it. If you shout toward a hard surface, the sound waves bounce off the surface. You may hear an echo as the sound waves return.

A sound that bounces off a hard surface is called an echo. You've probably heard an echo before. If you yell "Hello!" your vocal cords make sounds that travel away from you in all directions. If the sound waves hit a hard surface, they may bounce back to you. If the echo is strong enough, you hear yourself yelling "Hello!" after you have already yelled. If there is more than one hard surface, an echo may sound several times.

 Focus Skill **MAIN IDEA AND DETAILS** How does sound travel?

Describing Sound

If you hit a drum gently, the sound is soft. If you hit the drum harder, the sound is louder. **Volume** is a measure of the loudness of a sound. Volume is measured in units called decibels (dB). The softest sound a human can hear is 0 dB. A whisper is about 20 dB, a conversation is about 60 dB, and a thunderclap is about 120 dB! Sounds above 100 dB can cause pain and damage to a person's ears. That's why it's important to wear ear plugs or other ear protection around loud sounds.

You can also describe a sound as high or low. You know from listening to music that some sounds are higher than others. The **pitch** of a sound is how high or low it is.

Very loud sounds can damage your ears. Always pay attention to the volume of your stereo or portable music player.

Guitar players change the pitch of the sounds they produce by changing the length of the guitar's strings. Changing the length of a string affects how fast it vibrates. A shorter string vibrates faster. A longer string vibrates slower. The number of vibrations per second is the **frequency** of a sound. The frequency of the sound waves determines the pitch of the sound. A sound with a high frequency has a high pitch. A sound with a low frequency has a low pitch.

Focus Skill **MAIN IDEA AND DETAILS** How does changing the length of a guitar string affect its sound?

Guitar players press the frets to change the length of the vibrating strings. This alters their pitch.

Tuning pegs tighten or loosen strings. A string's tension also affects the frequency of its vibration.

Thinner strings vibrate faster and produce a higher-pitched sound. Thicker strings vibrate slower and produce a lower-pitched sound.

What Is Light?

You know that you need light to see. You probably also know that light, like sound, is not matter. Light is a form of energy.

The light that you see is just a small part of the electromagnetic spectrum. The electromagnetic spectrum includes waves of light that differ in their frequencies. The part of the spectrum you see is called *visible light*. Other types of electromagnetic waves include radio waves, microwaves, infrared waves, ultraviolet waves, and Xrays. Some of these types of electromagnetic waves have frequencies that are lower than visible light waves. Some have frequencies that are higher.

Fast Fact

A rainbow forms in the sky when sunlight passes through raindrops. Each raindrop acts as a prism, separating white light into its range of colors.

Certain wireless devices, such as this cell phone, receive and convey signals through radio waves. Radio waves have lower frequencies than visible light waves.

You see the shortest wavelength of visible light as violet. You see the longest wavelength as red.

You might think of sunlight as white, or having no color. Sunlight is actually made up of all the colors of the rainbow. A prism demonstrates the spectrum of visible light. When white light passes through a prism, the different length waves that make up visible light bend at slightly different angles and separate. Humans sense these separate wavelengths as different colors of light. You sense long wavelengths of visible light as red, and short wavelengths as violet. Between red and violet are all the colors of the rainbow.

 **MAIN IDEA AND DETAILS** What makes up the electromagnetic spectrum?

Light Waves

Light travels in waves. However, light waves are different from sound waves. Remember that sound waves travel through matter as compression waves. Light waves do not need a medium to travel, and they do not compress matter as they travel. Light waves move more like ocean waves. You may know that waves of water move up and down, but not forward or backward. This is similar to how light waves move.

In space, light travels about 300,000 km (186,000 mi) per second. It takes light from the sun about 8 minutes to travel the 150 million km (93 million mi) to Earth.

Since light waves do not need matter, light can travel through empty space. For example, the sun gives off visible light and other wavelengths of light from the electromagnetic spectrum. These waves travel through space to Earth.

Light waves pass through some materials but not others. Materials that allow light to pass through them are called **transparent**. Transparent materials include glass, water, and some plastics. Materials that allow only some light to pass through them are called **translucent**. Translucent materials include waxed paper, bubble wrap, and gauze. Materials that do not allow any light to pass through them are called **opaque**. Most materials are opaque—you can't see through them.

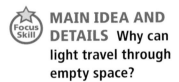 **MAIN IDEA AND DETAILS** Why can light travel through empty space?

Windows are made of glass, because glass is transparent.

Absorption and Reflection

When light energy strikes an object, some of it may be absorbed, some of it may bounce back, and some of it may pass through.

Absorption takes place when the light that hits an object is taken in. Objects of different colors absorb different amounts of light. Objects that are dark in color absorb the most light. However, objects don't absorb all of the light that hits them. Light that does not pass through an object is bounced back. The bouncing of light from a surface is called **reflection**. When light reflects off the surface of an object, it changes the direction it is traveling.

Light bounces off a mirror and creates an image. The image in a flat mirror is always upright, life sized, and left-to-right reversed.

> Light is reflected off a smooth, shiny surface at the same angle it hits the surface.

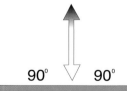
90° 90°

25° 25°

70° 70°

You see objects because light reflected from the surface of the object enters your eyes. A mirror is very smooth and shiny. A smooth surface reflects light in a predictable way. When light is reflected from a mirror, it is reflected at an angle equal to the angle at which it hits the mirror. When you stand in front of a mirror, you see a face just like yours. This is your reflection.

Most objects or surfaces are not as smooth and shiny as mirrors. Light that is reflected from surfaces that are dull or rough are reflected in many different directions. You don't see a reflection when light is reflected in different directions.

 MAIN IDEA AND DETAILS
What happens when light energy strikes the surface of an object?

Fast Fact

Did you know that the moon does not give off any light of its own? The moon appears to shine in the night sky because of the light it reflects from the sun.

13

Refraction

When light passes through some materials, it changes speed. The change in speed causes the light to bend. For example, when light passes from air to water or water to air, it bends. The bending of light as it moves from one material to another is called **refraction**.

When you look at an object through two different materials, the refraction of light changes the angle at which you see things. Looking at an object through both water and air, for example, can make a solid object look as if it's in two parts!

 MAIN IDEA AND DETAILS **How does the refraction of light affect the way you see things?**

The pencil looks like it's in two parts. Light traveling from the pencil to your eyes bends as it moves from the water, through the glass, and into the air.

The refraction of light is a useful property. Magnifying lenses are curved lenses that bend light to make objects appear larger.

Summary

Sound and light are forms of energy that travel in waves. Sound energy travels through matter in compression waves caused by vibrations. You can describe a sound by describing its volume, pitch, and frequency. Light energy travels through space and certain materials as electromagnetic waves. When light waves strike the surface of an object, they are absorbed, reflected, or refracted.

Glossary

absorption (ab•ZAWRP•shuhn) The taking in of light or sound energy by an object (4, 12)

frequency (FREE•kwuhn•see) The number of vibrations per second (7, 15)

opaque (oh•PAYK) Not allowing light to pass through (11)

pitch (PICH) How high or low a sound is (6, 7, 15)

reflection (rih•FLEK•shuhn) The bouncing of light off an object (12, 13, 15)

refraction (rih•FRAK•shuhn) The bending of light as it moves from one material to another (14, 15)

translucent (tranz•LOO•suhnt) Allowing only some light to pass through (11)

transparent (tranz•PAIR•uhnt) Allowing light to pass through (11)

vibration (vy•BRAY•shuhn) A back-and-forth movement of matter (2, 3, 4, 7, 15)

volume (vahl•YOOM) The loudness of a sound (6, 15)